Science of Electricity:

Volume 4

Wind Power Technologies Explained Simply

by Mark Fennell
© 2012

This book is part of the
Energy Technologies Explained Simply™ Series

Other Books in the Energy Technology Series

Renewable Energy Books
Introduction to Electrical Power
Hydropower Technologies Explained Simply
Solar Power Basic Concepts
Practical Considerations of Solar Power
Advanced Solar Cell Technologies
Wind Power Technology Explained Simply

Coal Power Books
Formation and Mining of Coal
Clean Coal Technologies
Mercury and Coal Power

Nuclear Power Books
Nuclear Power Meltdowns and Explosions
Health Hazards of Radioactive Decay
Radiation Measurements
Processes of Radioactive Decay and Storage of Nuclear Waste

Natural Gas Books
Natural Gas Basics
Extracting and Refining Natural Gas (includes Fracking)
Transportation, Storage, and Use of Natural Gas

Power Line and Grid Books
Introduction to the Transmission of Electrical Power
Power Lines
Underground Cables
Utility Operations and Quality Control
Power Grids Explained Simply

About the Book

This book explains all the important technology and practical tips related to wind power. Here you will learn everything you need to know to design, select, and install a wind turbine for your specific needs.

This book is designed as an overview for decision-makers at all levels, as a practical guide do-it-yourself types, and an easy read for curious citizens.

This book will also be a valuable reference work for students entering the growing workforce in wind power technologies, including the areas of design, manufacturing, and installation of wind turbines.

The first chapter explains all the basic concepts of wind power, including a discussion of the factors which affect the amount of power produced.

The second chapter discusses several practical details when installing wind turbines. Topics include: placement, height, orientation, storms, durability, and bird strikes. This chapter also discusses what to do when there is no wind.

The third chapter discusses blade design. Topics include: blade shape, number of blades, angle of attack, tip speed ratio, and basic aerodynamics. This chapter also discusses methods for calculating wind power from measured data.

The fourth chapter discusses offshore wind turbines. Topics include: storm resistance, placement of undersea cables, shipping lanes, and visual obstruction of turbines.

The final chapter explains additional wind power terms and concepts. Some of the concepts include: camber, lift force, stall, cut-out speed, and Reynolds Number.

About the Energy Technology Series

Purpose of this Series

The books in the *Energy Technologies* series are designed to educate citizens, students, and legislators on all aspects of energy technologies. The first books in the series focus on electrical power.

The books discuss many energy technologies, including: generators, turbines, power plants, power lines, and grids. The technologies for each type of power source (hydro, wind, solar, coal, nuclear, and natural gas) are discussed in detail. The books also discuss efficiency, safety, reliability, and health concerns for each energy technology.

The ultimate goal of the series is to enable the people to make informed decisions on practical energy questions. The secondary goal is to serve as introductory guides for students embarking on careers with energy technologies.

Taken altogether, the books in the series answer any question you are likely to have, such as:

- How can we increase the efficiency of solar cells?
- How do I select the size my solar array?
- What do I need to know when installing a wind turbine?
- How effective are the clean coal technologies?
- How can we prevent grid failures?
- Do power lines cause cancer?
- and many other energy technology questions...

Science of Electricity in Perspective

The subject of electrical power is of great importance to our communities, but is rarely taught. Public debate is frequent and passionate, but with too little understanding of the actual science. At best, an informed citizen knows only a few pieces. At worst, as it is for a great number of citizens, electricity is magic and myths are believed as scientific truth. It does not have to be that way. Any citizen, regardless of background, can know the technologies behind all aspects of electricity.

The books in this series solve that problem. These books educate the general public in all aspects of electrical power. Any person, regardless of background, can easily find the answer to his energy question in one of these books.

Specific Goals

There are numerous technologies described in these books. Yet for each technology I sought out the answers to the following questions:

1. How does the technology work?
2. What are the advantages and disadvantages?
3. What is the efficiency? How can the efficiency be improved?
4. What is the environmental impact? How can it be improved?
5. What are the safety hazards, and how can they be reduced?
6. What are the most important practical tips?
7. What facts comprise the most important data?

Technical Discussions Explained Simply

The books in the series must necessarily be technical to some degree. Electricity is a practical technology, and therefore we must understand the technical aspects if we want to make wise decisions. Yet the discussions in this book are always aimed at the citizen or policy maker.

The books in this series explain the principles of electricity as simply as possible, using ordinary English (no engineering jargon), and highlighting the most important points of each technology. Main concepts and facts are emphasized with the use of lists, tables, diagrams, and summaries.

I do not expect any reader to have a background in science, yet I offer enough facts and details so that the reader can have an accurate understanding of all related technologies. I provide enough technical details and enough data for the reader to make informed decisions.

Conclusion

For all the reasons above, I offer this series of books. My goal is to inform you on the basic concepts of all the technologies and all of the issues related to electricity so that you can make realistic decisions.

Remember that there are no perfect solutions, there are only choices. I hope that this series of books will assist you in making those choices for your community.

Mark Fennell

Accuracy and Technical Depth

Objectivity

I have tried my best to be as objective as possible. Whereas many other authors of energy books have an agenda, I have no desire to promote one industry over another. I have no desire to promote one technical solution over another. In this endeavor, I have tried to be an objective scientist.

Accuracy of Data and Summaries

I never relied solely on the conclusions of other researchers. Instead, I performed many other tasks to ensure that all conclusions were accurate. I examined primary data whenever possible. I have read the fine print on how research was obtained.

I have also checked the accuracy of the conclusions written by other researchers, most commonly by finding at least three distinct sources for each fact. In addition, I performed my own calculations numerous times to prove (or disprove) conclusions and final values in other reports. It is only after such rigorous investigations that I created data tables and wrote summaries for these books.

Limited Mathematics

The books must also use math from time to time. For example, efficiency is a statement of a specific amount, and therefore the discussion of efficiency requires the use of equations. Other issues such as power loss, health hazards, environmental concerns, and quality control are also statements of amounts and also require calculations. Therefore some equations are necessary to know, even for the non-scientist.

I also provide examples of calculations so that readers can become more comfortable with using the equation themselves.

However, I want to emphasize that I focus on concepts not on the mathematics. I provide equations only when it is necessary for the citizen or student to be familiar with these equations.

Table of Contents

4.1
Wind Power Basics

Introduction

Wind power is a new twist on an old technology and is growing in popularity. Wind turbines are relatively simple to operate and very low maintenance. Wind turbines are environmentally friendly and can provide power for remote areas. However, wind is variable and therefore cannot be relied on for steady power.

List of topics for this chapter
1. Brief Description of Wind Power
2. Quantity of Wind Power
3. Density of the Air
4. Area Swept by the Rotor of the Wind Turbine
5. Velocity of the Wind

Brief Description of Wind Power

Wind is already an item in a state of flow so there is nothing to convert. The wind hits the blades of the turbine, which causes the turbine to rotate. The rotating turbine operates the generator, which then creates the electricity.

A wind turbine is essentially a modern, efficient windmill. We are familiar with windmills, for they have been used for centuries on farms to pump water. (Figure 4.1) The wind turbine looks very similar (Figure 4.2). However, there are important technical differences between the terms "windmill" and "wind turbine." A windmill performs a mechanical function, mostly for pumping water. A wind turbine creates electricity.

Quantity of Wind Power

Wind Power Equation

The amount of power from a wind turbine depends on: 1) density of the air, 2) area swept by the blades of the wind turbine, and 3) velocity of the wind. Of these factors, wind velocity is the most important. Wind power can be approximated by this simple equation: Power = ½ x Density of air x Area swept by turbine x (Velocity of wind)³. This power equation can be abbreviated to: Power = ½ (D)(A)(V)³.

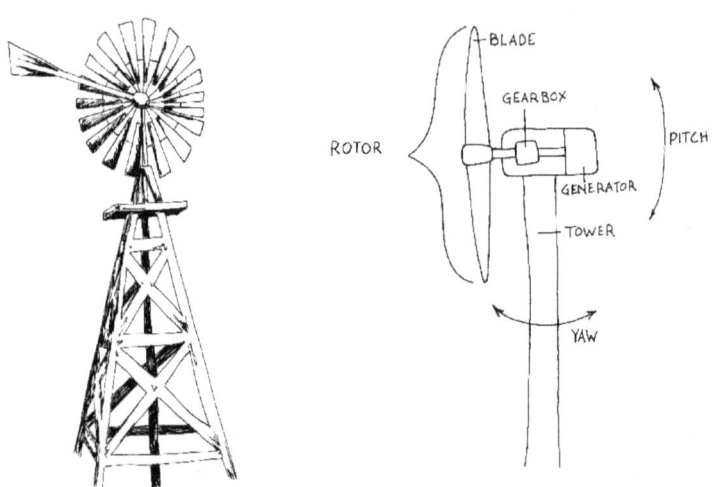

Fig. 4.1 Classic Windmill   Fig. 4.2 Basic Wind Turbine

Wind Power Units

Many sets of units can be used in the wind power equation. One commonly used set of units is the following:

1. Density of air: measured in kilograms per cubic meter (kg/m³)
2. Area swept by turbine: measured in square meters (m²)
3. Velocity: measured in meters per second (m/s)
4. Power: final result is in Watts

The units of the wind power equation would then be:

Watts = ½ (D)(A)(V)³ = .5 x kg/m³ x m² x (m/s)³

Note that the final units become = .5 x kg x m²/s³, and recall from Part 1 that the unit of kg x m²/s³ is the same as a watt.

Density of the Air

The concept of greater density means that more molecules are bunched together. Regarding wind turbines, when air is denser we get more molecules hitting the turbine blades at the same time. This results in a greater push. Therefore, air that is denser will produce more power than air that is less dense.

The density of the air is related to temperature: cold air is denser than warm air. Therefore, cold air gives us more power. This translates into these practical points:

    a. Winter winds provide more power than summer winds.

    b. Colder areas can get more power than warmer areas.

Area Swept by the Rotor

The area swept by most wind turbines is easy to imagine. For most wind turbines, the area swept is the area of a circle. The area of any circle is calculated by: Area = Π (radius)^2. For a wind turbine, the radius is the length of the blade. Therefore, increasing the blade length, even by a short amount, has a significant effect on the amount of power produced.

Velocity of the Wind

Velocity of the wind is the most important factor, yet it is also the most complex. We can see that velocity is important from the power equation. The velocity term is cubed, therefore any increase in velocity, even a small one, will make a significant increase in power. Therefore, we want to do whatever we can to get the greatest wind velocities to our turbines. However, because the velocity of wind is variable, our calculations become more complex. Wind speeds vary; unlike other power sources, we have no control over the flow of wind. We will look at the ramifications, and the practical points, in subsequent chapters.

Chapter Summary

1. The power from a wind turbine depends primarily on: Density of the air, the Area swept by the blades of the wind turbine, and the Velocity of the wind.

2. Power from a wind turbine can be calculated from the equation:

   Power $= \frac{1}{2} (D)(A)(V)^3$

3. Cold air is denser than warm air. Therefore we can create more power when the wind is cold. This translates into two concepts: a) we can create more power during the winter than during the summer, and b) colder regions can create more power than warmer regions.

4. A longer blade sweeps a larger area, and therefore a longer blade provides more power.

5. Velocity is the most important factor in creating power from wind. However, velocity is also a complex factor.

4.2
Practical Considerations of Wind Power Operation

Introduction

Wind power can be very effective. However, if we are to get the most from our wind turbines then we must understand a few practical considerations.

List of topics for this chapter
1. Placement
2. Height and Obstructions
3. Orientation
4. Blades
5. Cold Weather, High Winds, Storms
6. Lightning
7. No Wind: What to Do
8. Birds
9. Durability
10. Radio Interference

Placement

We want to put wind turbines where we can get the most wind. Prime places include mountaintops, flat plains, and ocean coasts. Mountain tops and flat plains are both ideal locations because the winds are strong and because there are few obstructions. Most of the wind turbines in this country are placed on mountain tops or on flat plains. Offshore wind turbines deserve special attention. The oceans provide a great location for wind turbines. However, there are additional design considerations. We will discuss the practical factors for offshore turbines in a later chapter.

Obstructions and Height of Turbine

It is important that we do not have any obstructions. Obstructions such as trees break up the wind before the wind gets to the turbine. If the wind is broken up then the force is less, which results in less power. Turbines must be built tall enough to not be affected by nearby obstructions. The most common obstructions are trees and houses. In practical terms, wind turbines must be taller than all trees and buildings within 150 feet of the turbine.

Orientation (Figure 4.2)

The turbine must face the wind as often as possible. Because wind directions change often, it is important to have some mechanism which can turn the rotors of the turbine to meet the wind. Traditional windmills use tail vanes. The wind pushes the tail vane and then the rotor turns to meet the wind. Many small wind turbines also use this tail vane system. Larger wind turbines, such as those used in large wind farms, often have mechanical devices built in which allow them to be turned from a control room.

Changing the orientation of the rotor involves changing the pitch and the yaw. Pitch is the up and down movement. Yaw is the left to right movement. When we want the blades to face the wind we adjust the yaw. When major storms occur, and are worried about damage to our turbine, we adjust the pitch.

Blades

The number of blades needed for a turbine depends on the average wind speed in a particular location. If the average wind speed is mild then more blades are required (5 or more). However, if the location has high wind speeds, then a few blades (2 or 3) will be sufficient.

The traditional windmill rotor is called "multiblade." Windmills require multiblade rotors because windmills are often used in locations of mild winds. However, wind *turbines* are usually placed in areas of high velocity winds and therefore a rotor with 2–3 blades works best.

The material of the turbine blade must withstand rain, flying debris, and the wear from repetitive rotation. Turbine blades are usually made from composite materials such as fiberglass.

Metal should not be used for a variety of reasons, including:

a. Repeated rotation quickly causes metal fatigue

b. Metal interferes with radio wave communication

c. Metal rusts, especially in sea air (a problem if near the coast).

Blade design is important because the efficiency of the turbine is directly related to the design of the blades. The concepts of blade design are discussed in detail in a separate chapter. However, we can cite a few of the more important points here:

1. Blade size: long and thin blades provide most power.

2. Shape: an aerodynamic shape provides more power than just a flat blade.

3. Angle: The blade must face the wind at a specific angle ("angle of attack"). The optimum angle is computed so as to get the maximum power.

Cold Weather, High Winds, Storms

We like cold weather because the air is denser and therefore we get more power. However, we don't like the weather so cold that ice builds or that machinery freezes. Many turbines that exist in very cold climates have heaters built into the system. These heaters prevent the oil from freezing, and thus prevent the machinery from stalling.

We like strong winds because higher velocity winds create more power. However, we don't want winds so strong that our turbine becomes damaged. In order to prevent a wind turbine from being damaged from very high winds, most turbines have an "overspeed control." There are two main methods of overspeed control:

1. The first method shuts down the turbine automatically when the wind speed reaches a certain maximum level. The speed at which the turbine shuts off is the "cut-out" speed.

2. The second method of the overspeed control is to turn the rotor in such a way so as to get less force from the wind. Usually this means adjusting the pitch.

Lightning

Wind turbines are often the tallest objects in their area (so that there are no obstructions). However, the disadvantage of being the tallest object is that lightning will strike the turbine rather than anything else. To prevent damage from lightning we can do the following:

a. Place a lightning rod on top of the turbine tower. This lightning rod is connected to a wire which leads directly into the ground.

b. Use composite materials for the blades and the tower, rather than metal or pure wood. (Metal will conduct the electricity of lightning. Wood will burn when hit by lightning.)

c. Place an electrical insulator on the tip of the blades. A blade with an electrical insulator is less likely to attract lightning.

What to Do for Times of No Wind

Introduction

Wind speed is variable, and often there is no wind at all. If a community wants to have electrical power on a regular basis then they must find a way around this obstacle. The main methods include the following:

1. Batteries
2. Conjunction with hydropower
3. Small hydropower on site
4. Wind used only as supplement to other power

Batteries

Batteries as storage for wind power are useful for small sites such as for a single home. However, batteries are not practical for wind farms which provide power to a large community.

Conjunction with Hydropower

Hydropower and wind power can co-exist nicely. This means that we can get two sources of electricity in the same general area. In this system, wind turbines provide most of the power and the reservoir will supply the rest. When the wind is slow or stops completely, the sluice gates of the dam can be opened up, allowing the creation of additional power.

Small Hydropower on Site

Many wind farms have their own mini hydroelectric power plant. The same scenario exists as above, where wind power is the primary source and hydropower only supplements power when needed. However, this hydropower plant is a much smaller version. In this system, part of the power from the wind is used to pump water to a small reservoir. The rest of the wind power is used to create the actual electricity. Then, when the wind is slow or non-existent, the mini hydro-plant is used to supply the remaining power.

Wind Used only as Supplement to Other Power

The best method of using wind power is to have wind power as the *secondary source*, not as the primary. Wind power can rarely be used on its own. Even if we had enough wind turbines, the variance in the wind would leave us with an unpredictable generation of electricity. Therefore, wind power is best used as a supplement to other main forms of power.

In this system the primary power source would be coal, nuclear, or hydro. There can still be a wind farm, even a very large one, but the wind farm is not relied on as the primary source. This wind farm will produce as much wind power as possible. Traditional sources such as coal provide the rest. Then, as the wind slows or stops, the main power plant increases its rate of electrical generation.

Birds

Birds like to travel on the same wind currents that are ideal for turbines, and therefore the birds can sometimes become hurt in the process.

The issue of birds and wind turbines has only recently been recognized. Factors that must be studied related to this issue include: types of birds, types of turbines, geography, climate, and seasons. Note that factors vary so much at each wind power location that each location must do its own study.

Some of the best practices known today for preventing injuries to birds include:

1. <u>Avoid building turbines along major migratory paths</u>
   Some wind patterns are used by birds more than others. If numerous birds use this path on regular basis then there are greater odds for birds to be injured by turbines on the path. Instead, place the turbines in those windy areas which are not migratory paths. (There are plenty of these locations available)
   It is true that some birds will use these winds for flight, but not as many, and therefore the odds of injuring a bird are far lower.

2. <u>Clear the territory of any of the main sources of food</u>
   If there are is no food for the birds, the birds will not come here. Therefore clear the area of rodents, insects, and other such food which is tempting to the birds. With no food, the birds will have no incentive to come, and therefore will not be injured by the blades.

3. <u>Construct turbines that are not conducive to perching</u>
   Birds like to perch on top of the turbine gearbox, often making a nest while the blades are not moving. Because the birds make this turbine their home, they will come and go...and if the turbine blades are operating the bird may become injured. To prevent this event, simply place spikes on top of the gearbox, which will discourage the birds from resting there.

4. <u>Use turbine blades with lower tip speed ratios</u>
   Lower tip speed ratios will produce a slightly slower rotational speed. Therefore, design the blades with lower tip speed ratios which will then reduce the seriousness of the injury should a bird be hit by the blade.

Durability

There is a factor which some people call "relative mass" which seems to provide an indicator of durability. Relative mass means: mass of the turbine relative to the area swept by the rotors. A higher value of relative mass (greater mass as compared to the area swept) seems to have greater durability.

For example, if the mass of the turbine is 40 kg, and the area swept is 10 m², then the relative mass is 40/10 or 4/1. Then we compare turbines. A turbine with a relative mass of 4/1 is more durable than a turbine with a relative mass of 3/1.

Radio Interference

Wind turbines have the potential to interfere with radio waves. Radio signals can become distorted by wind turbines. The most significant interference comes from cell phones and AM radio. Owners of wind turbines and owners of communication systems must work together before building towers.

Communication systems operating in the microwave range of wavelengths are the most susceptible to interference from wind turbines. These communication systems include cell phones and point-to-point microwave dish systems.

Cellular phones are the most likely communication devices to be an issue for wind turbines. Some of the reasons include: a) The cell phone towers are approximately the same height as most wind turbines, b) the short wavelengths of the cell phone can be affected by small wind turbines, and c) both cell phone towers and wind turbines are placed throughout the countryside.

Chapter Summary

1. The main practical issues in wind power operation are: placement, obstructions, orientation, blades, storms, supplementing wind power, and durability.

2. The best locations for wind turbines are mountain tops, flat plains, and ocean coasts.

3. Turbines must be built taller than any nearby obstructions.

4. Turbine rotors must face the wind. Turbines must have a mechanism for turning the rotor to the meet the changing direction of the wind.

5. Turbine rotors are moved in two directions. Pitch is the up and down movement; yaw is the left to right movement.

6. The number of blades needed for a turbine depends on the average wind speed in a particular location.

7. Long and thin blades provide the most power.

8. An aerodynamic shape provides more power than just a flat blade.

9. We like cold weather because the air is denser and therefore we get more power, however we don't want machinery to freeze. In order to prevent freezing some turbines are equipped with heaters.

10. We like strong winds because higher velocity winds create more power, however we don't want winds so strong that our turbine becomes damaged. There are two methods to deal with winds that are too strong: a) the turbine is automatically turned off, or b) the turbine rotor is turned in such a way so as to get less force from the wind.

11. Wind speed is variable. If a community wants constant power this issue must be taken into account. Methods include: batteries, wind used only as supplement to other power, or small hydropower on site.

12. The relative mass provides an indicator of durability. Relative mass is the mass of the turbine relative to the area swept by the rotors. The higher value (greater mass as compared to the area swept) has greater durability.

13. Wind turbines can cause radio wave interference. Communication systems operating in the microwave range of wavelengths are the most susceptible to interference from wind turbines. These communication systems include cell phones and point-to-point microwave dish systems.

4.3
Blade Design

Introduction

There are many practical factors in blade design. A good blade design can make a significant difference in the amount of power produced.

<u>List of topics for this chapter</u>
1. Area: Blade Size and General Shape
2. Area versus Velocity
3. Velocity, Area, and Final General Blade Shape
4. Blades Fixed at an Angle: Angle of Attack
5. Basic Operation the Wind Turbine
6. Aerodynamics
7. Number of Blades
8. Horizontal Axis vs. Vertical Axis
9. Calculating Wind Velocity

Area: Blade Size and General Shape

We noted from the power equation that one of the key factors is area swept by the blades. A rotating turbine sweeps in a circle, the area of which is: $A = (\Pi)r^2$. The blade is in essence the radius of the circle that is swept by the turbine. Therefore, a longer blade sweeps a larger area, which then creates more power. If we have a wider radius (a wider blade) then we would sweep more air.

The volume of air, as related to width and length of the blade, can be calculated by: $A=(\Pi)Wr^2$, where "W" is the width and "r" is the length of the blade. We can see that the larger amount for width we put into the equation, the more area we will cover. Therefore, in general, a wider turbine blade provides more power than a thin turbine blade. The classic Dutch windmills used for centuries did indeed have blades that were long and wide. For many years, this was the best design.

Area versus Velocity

Introduction

The large area blades worked well enough for many years. However, velocity eventually became a more important factor than the area. This is for two reasons: 1) the purpose of the wind turbine differs from the purpose of windmill, and 2) velocity has a much greater effect on creating power than the area does.

Wind turbines differ from windmills in one important way: windmills were used only for mechanical functions, such as milling and pumping water. The speed of the windmill was not a factor. Consistent operation was far more important than how fast the operation was done.

In contrast, a wind turbine is used to create power. Because power is a rate, the speed of rotation is very important. Therefore a turbine which rotates faster will produce more power.

Although this concept can exist for any turbine, it is most important in the arena of wind turbines. Therefore, it is very important that we do everything we can which will increase the rotational speed of wind turbines.

Furthermore, the velocity has a much greater effect on creation of power than the area, particularly for wind turbines. Recall that the power equation is: Power = $\frac{1}{2}$ (D)(A)(V)^3. Because the velocity term is cubed, an increase in velocity will have a much greater effect on the power than an increase in area. Therefore, designing for velocity is far more important that designing for area.

Velocity, Area, and Final General Blade Shape

Now that we realize that the velocity factor over-rules the area factor in blade design, we can take a second look at the size of the blades. In general, thinner blades provide better velocity to the turbine than wider blades. Remember, we are not so much concerned with the force of the wind on the blade, but rather we are more interested in rotating the turbine at a faster speed. Thinner blades help us rotate the turbine faster than wider blades, and therefore thinner blades are used most often.

In addition, we like to do what we can for increasing area. Longer blades cover more area than shorter blades, even if the blades are thin. Therefore, longer blades are used more often than shorter blades.

In total, due to considerations of velocity and area, the blade shape that provides the most power is the general shape of thin (creates faster rotation) and long (covers more area).

Angle of Attack: Blades Must Be Fixed at an Angle

At this point we want to make a distinction between the rotor and the blades. We want to face our *rotor* directly into the wind. However, we do not want our *blades* to face directly into the wind. If we faced our blades directly into the wind then the blades would not move. This would just push on the blades, but not rotate the windmill.

Therefore, we desire the blades of our turbine to be at an angle. In order to get our wind turbine to turn, we need to have our *rotor* face the wind directly, but we need our *blades* tilted slightly with respect to the wind. When the blade is tilted slightly in relation to the wind, it is then that the blade can be pushed in the rotation that we desire.

The "angle of attack" is this slight angle which we desire. Specifically, the angle of attack is defined as the angle of the blade relative to the direction of the wind (Figure 4.4, 4.6). The smaller the angle of attack, the faster the rotor will turn. The larger the angle of attack, the slower the rotor will turn. We want the rotor to turn faster, and therefore our angle of attack should be small. Thus, although we must have our blades tilted in relation to the wind, the blades should be tilted by only a small amount.

Basic Operation of the Turbine

Before proceeding to advanced blade design subjects such as aerodynamics we must first understand how wind causes the turbine to rotate. (Figure 4.4) A turbine blade usually has two flat sides, which can be called the "top" and the "bottom." The "bottom" of the blade is the side of the blade that faces almost straight into the wind. The "top" of the blade is the side that faces away from the wind. The bottom side is turned slightly so that the wind will hit the blade at an angle, and thus push the blade sideways. Then, because the blades are attached to a rotor, this sideways push causes the blade to move in a rotation. Having several of these blades on the same rotor allows the wind to hit a blade at any time. This result is a continuous rotation of the turbine.

Aerodynamics

Introduction

In addition to the basic rotation of the wind turbine we can also consider the aerodynamics. Special aerodynamic design of wind turbine blades can increase the velocity, which will result in increase in power.

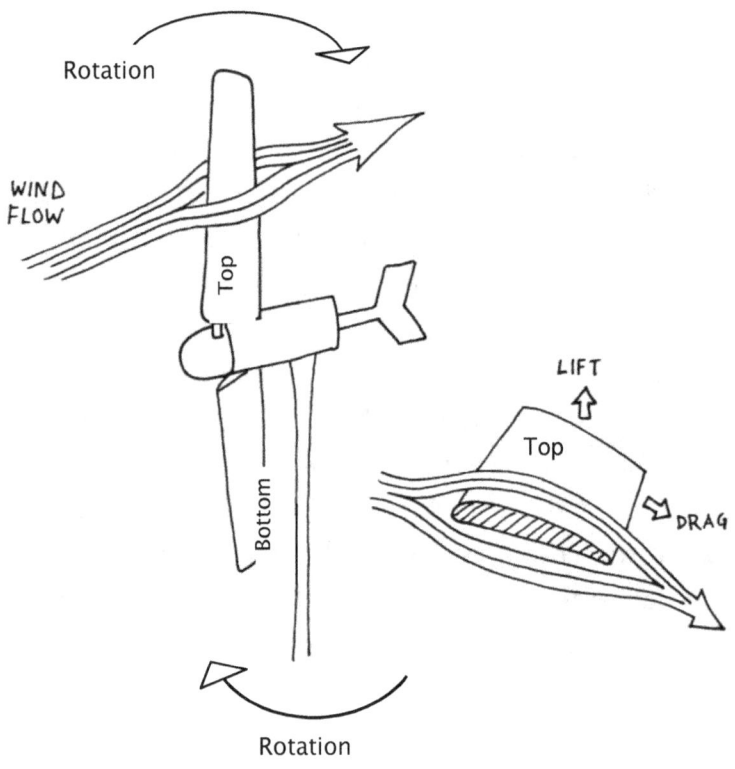

Figure 4.4: top of blade, bottom of blade,
angle of attack, rotation, lift, drag

Tapered Blade Thickness (Figure 4.4, 4.6)

Most modern turbine blades have a tapered thickness (Figure 4.4). This makes the blades look much like the wing of an airplane. The reason for this tapered thickness is to improve the aerodynamics. The blade is thick in the front and progressively thinner in the back. Due to the tapered thickness of the blade, the wind turbine actually spins several times faster than the speed of the wind.

<u>Tip Speed Ratio</u>

Due to the aerodynamic shape of the modern turbine blade, the rotation of the wind turbine spins faster than the wind speed. The "tip speed ratio" expresses this relationship of the rotational speed vs. the wind speed. Specifically, the tip speed ratio is defined as: the speed of the tip of the blade as compared to the speed of the wind. (Hence: "Tip Speed Ratio")

Note that the tip is used rather than any other part of the blade because the tip often travels faster than any other part of the blade. Furthermore, the tip is often designed with additional aerodynamic geometry, often making the tip even more aerodynamic than the rest of the blade.

We want our wind turbine to rotate faster, for that will create more power. This means that we want to increase our tip speed ratios. A wind turbine should have a tip speed ratio of at least 5 to 1, ideally as high as 10 to 1.

Number of Blades

The number of blades we need depends on how fast the wind usually travels in that area. Many sources say that 3 blades are enough. However do not be misled, the number of blades depends on where you live.

In general, we want to always have at least one blade meeting the wind at all times. Imagine if we had only one blade on our turbine. The force of the wind might not be enough to push the blade all the way around again. If this happened then the windmill would be useless.

Similarly, two or three blades may not be enough – the force of the wind might not be able to push the one blade far enough for the next blade to come the wind impact location. That is why traditional windmills have numerous blades. Even if the wind is very light, there is another blade right behind to catch more of the wind.

Hence, with a multi-blade windmill, the windmill can still rotate even when the winds are light. Similarly, with a multi-blade wind turbine, the wind turbine can still rotate even when the winds are light.

However, if you live in an area which gets gusty winds most of the time then you can get by with fewer blades. When authors of various reports state that the three-blade turbine is good, these authors are really giving a hidden assumption. Three blade turbines are good only if you live in windy areas. Yet, many authors assume that you wouldn't choose to use wind power unless you *did* live in a windy area.

Assuming you live in a windy area, where winds are brisk most of the year, then a three blade turbine will be fine. However, if you just want a few wind turbines on your property to supplement your electricity supply, and you don't live in a location of constant wind, then you should think about having more blades.

Horizontal Axis vs. Vertical Axis

The classic windmill is horizontal axis. The most common wind turbines are horizontal axis turbines. When you think of a windmill or wind turbine, it is usually a horizontal axis version. In contrast, vertical axis turbines look like eggbeaters. These were popular briefly in the 1970s and 1980s, however these vertical designs have not proven to be any better than the horizontal axis. Nor are the vertical axis turbines very practical. Therefore, because vertical axis turbines are not useful we will not discuss them.

Calculating Wind Velocity

Introduction

Wind velocity is the most important factor in getting electrical power from a wind turbine. The velocity term is in fact cubed $(V)^3$. However, we have a difficulty in the calculation because the wind speed varies. With other power sources such as coal or hydro we supply the power and therefore we can keep the velocity steady. In contrast, with wind power we have no control over the velocity of the wind. Therefore we have two issues regarding calculation of the wind velocity term:

1. The velocity factor is cubed in the power equation.
2. However, the wind velocity varies, which makes it more difficult to get a number to put into the final equation.

What should we do? In this section we discuss the options.

Algebra Methods

We have two math factors here: 1) averages 2) cubes. Averages are necessary because wind speed is not consistent. Therefore it makes most sense to take an average wind speed. This means, for example, taking wind speed data every day for a month, then dividing by the number of days. However, in addition to the averages we have cubes. Recall that the equation for power of a wind turbine has the term $(V)^3$.

There are two ways to combine an average and cubing: "Averages of the Cubes" and "Cube of Averages." Depending on which way you do it you get two different results. An important concept to note here is that you get more power from the days of high velocity winds than what the average velocity value alone would indicate. Also note that some wind power producers prefer calculating this velocity using the Averages of the Cubed.

A. Averages of the Cubes
1) Do the cube of each term. 2) Take the average of the total.

B. Cube of the Averages
1) Take the average of all terms. 2) Do the cube of the total.

Examples

I will do one example of each, using the same wind speeds and same turbines. Suppose that I take wind speed readings once a day, every day, for a week. I then have 7 values for wind speed:

Mon = 2 m/s   Tues = 2 m/s   Wed = 3 m/s   Thurs = 5 m/s
Fri = 5 m/s   Sat = 5 m/s   Sun = 6 m/s

Example using "Averages of the Cubed"
$(V)^3$ term total = $[(2)^3 + (2)^3 + (3)^3 + (5)^3 + (5)^3 + (5)^3 + (6)^3]/7$
$= 90.5$ m/s

Example using "Cube of Averages" (average the speeds, then cube)
$(V)^3$ term total = $([2 + 2 + 3 + 5 + 5 + 5 + 6]/7)^3$ = 64.0 m/s

Therefore in this example we can see two things: a) you get different results based on which method you use to calculate, and b) the average of the cubes is higher than the cube of the averages.

Because higher wind velocities have a greater effect than a simple average indicates, many people who work with wind power prefer to calculate the velocity term using the "average of the cubes" method. This method tends to get closer to the value of actual power produced.

Wind data available and graphs

There are many sources for getting wind data (tables and graphs) for your area. Sources for wind data and graphs include:

1. Regional airports
2. Wind power associations (there is usually one in each state)
3. The National Climatic Data Center in Asheville, North Carolina.

Note that the National Climatic Data Center offers data in many forms on-line, at www.ncdc.noaa.gov.

Chapter Summary

1. Blade design is a very important factor. If we want to create the most power from a wind turbine, then we must have the most efficient blade design possible.

2. Regarding area swept and blade size:
   • A longer blade provides more power than a shorter blade.
   • A wider blade provides more power than a thinner blade.

3. Velocity is a more important factor than area: We want a wind turbine to rotate as fast as possible. A faster rotating turbine will create more power. Furthermore, the velocity term in the power equation is cubed.

4. Blade shape: Taking into account both velocity and area, the optimum blade shape is thin and long.

5. The blade must be tilted slightly in order to be pushed properly by the wind. The angle of blade tilt relative to the wind is the "angle of attack." The smaller the angle of attack, the faster the rotor will turn.

6. The turbine blade has a tapered thickness much like the wing of an airplane. The purpose of this is to improve the aerodynamics of the blade. The net result is that the turbine will rotate faster than with a flat blade and the turbine will create more power.

7. The tip speed ratio is the speed of the tip of the blade as compared to the speed of the wind. We want large tip speed ratios, between 5 and 10.

8. The number of blades we need depends on how fast the wind usually travels in that area. If you live in a windy area where winds are brisk most of the year then a three blade turbine will be fine. However, if you don't live in a location of constant wind then you should think about having more blades.

9. There are two methods for calculating the wind velocity term in the power equation: Averages of the Cubes, or Cubes of the Averages. Some wind power producers prefer the Averages of the Cubes.

10. "Averages of the Cubes"
    a) Do the cube of each term. b) Take the average of the total.

11. "Cube of Averages"
    a) Take the average of all terms. b) Do the cube of the total.

12. Sources for wind data and graphs include regional airports, wind power associations, and the National Climatic Data Center.

4.4
Offshore Wind Turbines

Introduction

The oceans provide a great location for wind turbines. Offshore wind turbines have several advantages, including: the winds coming off the ocean can be quite strong, offshore wind turbines can be built much larger than traditional wind turbines, and offshore wind turbines can create large amounts of power.

However, there are technological considerations for placing turbines offshore that differ from the considerations for installing wind turbines on land. Some of the more important considerations include:

- Access to equipment is difficult and thus maintenance is limited
- The turbine must resist the force of tides and storms
- Ocean water conducts electricity easily
- Power cables can become unburied or broken by weather or ships

List of topics for this chapter
1. Technological Considerations for Offshore Turbines
2. Where Offshore Turbines are Best Placed
3. How Far Out to Place Offshore Turbines
4. Shipping Lanes and Offshore Turbines
5. Placement of Undersea Cables
6. Visual Obstruction of Turbines

Technological Considerations for Offshore Turbines

Overview

The ocean creates additional technological considerations for wind turbines which are not seen on land. The main issues include: limited maintenance, damage from tides and storms, oceans conducting electricity, and cables can be easily damaged.

Limited Maintenance

Offshore wind turbines are difficult to maintain and repair. Access to the turbine, transformer, and cables can be difficult. Therefore, if offshore wind turbines are to be used then they should be as maintenance free as possible.

In order to prevent corrosion, every piece of the offshore turbine is coated with a non-corrosive protective layer. Many recent offshore turbines are also designed with features that make access for maintenance easier.

Resisting the Force of Tides and Storms

Any turbine fixed offshore must resist constant forces from tides and winds. (These forces are sometimes referred to as "water-wave" actions.) The base, the tower, the rotor, and the blades of the turbine must be designed to withstand those forces. Note that installing a sturdy foundation is usually the greatest cost of an offshore turbine.

Ocean Water Conducts Electricity Easily

Placing high voltage cables under water can be a dangerous decision. Ocean water has a high concentration of salt, and therefore the ocean water will conduct electricity very easily. If there is a break in the power line the risk of electric shock to swimmers can be significant. Furthermore, in order to transmit the power from the ocean to the coast, transformers are usually built on platforms in the ocean. These transformers boost the voltage in order to reduce power loss; however, the resulting voltage is now at dangerous amounts. In order to eliminate these dangers, best practices include:

a. Place turbines away from recreational areas

b. Choose cable designs which resist damage

c. Place the transformer (which increases the voltage) on the shore rather than on a platform in the water.

d. Use lower voltages to transmit electricity. Note that this might require placing the wind turbine closer to the shore.

Where Offshore Turbines are Best Placed

The best placement of offshore turbines will be related to the local weather, as well as how the area is used by local people. These concepts can be summarized as follows: 1) Higher velocity winds produce greater amounts of power. Therefore, any location with high velocity winds is ideal for offshore wind turbines. 2) Colder weather produces greater amounts of power. Therefore, coasts where the temperature is colder are ideal for offshore wind turbines. 3) Natural hazards will deter visitors. Therefore, any area which is unappealing to people will be conducive to building larger turbines and using greater voltage power lines.

Note that major fishing areas, swimming areas, and shipping lanes should be avoided. Anchors and fishing gear and damage cables, and sea water will conduct the electricity, resulting in major shocks to swimmers and wildlife.

How Far Out to Place Offshore Turbines

In general, the further out the wind turbine is from the coast the more power we can create. Given the technology and cost of installation today, the deepest we can build an offshore turbine is about 30 meters (approximately 100 feet). This depth generally places the turbine out at a maximum of 10 km (approximately 6 miles) from the shore.

However, there are complications as well. The further out the turbine is placed, the deeper the water. Building a steady base for the turbine is the most expensive step when installing a turbine, and the cost increases with the depth of the water. The longer distance also means that a longer cable is needed to reach the shore. The longer the cable, the more expensive it is to install. In addition, with a longer the cable there are more potential locations for the cable to become damaged.

Shipping Lanes and Offshore Turbines

Placement of turbines should also not interfere with shipping lanes and fishing areas. Anchors and fishing nets can easily drag into the ground and unbury the cables. After the cables are unburied, the next anchor to come along can easily snap the cables. We must also prevent wind turbines from being hit by ships. We want to avoid damage to either the ship or the turbine.

The best thing to do is avoid putting wind turbines in any shipping lane. However, no matter where we place the turbine that section of water will have some boats. Night and fog can prevent a boat from seeing even the largest of turbines. Therefore all offshore turbines need beacons which warn boats of the turbine. All of the details regarding shipping lanes and beacons can be worked out with the Coast Guard.

Placement of Undersea Cables

Many of the issues discussed previously are important considerations for laying cables under the ocean. The cables carry electricity, often at high voltages. Therefore it is very important that these cables do not become damaged. Furthermore, cables under water are difficult to access, making inspection and repair enormously difficult.

Note that laying the cables is one of the major costs for an offshore turbine (second only to installing the foundation for the turbine). Best practices for laying electrical cables under the ocean include the following:

a. The cables can be damaged most easily by anchors and fishing. Therefore it is best to avoid heavy fishing areas.

b. Tides can unbury the cables, leaving them exposed to a number of potential hazards. Therefore the cables must be buried deep enough to not be uncovered for years. Undersea cables must be buried a minimum of 3 meters below the ground.

c. Rocks can damage cables. Therefore cables may need to be encased in a protective layer where the ocean is rocky.

Visual Obstruction of Turbines

One of the objections to offshore turbines is how the turbines will look from shore. Offshore turbines can be as tall as 400 feet, so this objection is not trivial. However, any turbine placed a mile or more out to sea will barely be noticeable.

How big will an offshore wind turbine appear to the people on the coast? That answer depends on two factors: 1) the height of the turbine and 2) the distance which the turbine is placed from the land.

The best method for measuring how much of our visible space is taken by a wind turbine is to measure the "visual angle." The visual angle is measured from the horizon to the top of the turbine. A larger visual angle means that more of the wind turbine obstructs our view.

Note that a taller turbine will take up a larger visual angle than a smaller turbine, when both are situated at the same location. Also note that a turbine that is closer to the shore will have larger visual angle than a turbine of the same height which is placed further out. Detailed information for the visual obstruction of offshore turbines can be found in the appendix.

Chapter Summary

1. Offshore wind turbines can get a great amount of power from the ocean winds. However, there are additional design factors to consider.

2. Offshore wind turbines have several advantages:
    a. The winds coming off the ocean can be quite strong.
    b. Offshore wind turbines can be built much larger than traditional wind turbines.
    c. Offshore wind turbines can create large amounts of power.

3. The ocean creates additional technological considerations for wind turbines which do not exist for wind turbines on land. In general, the various issues exist because:
    a. Access to equipment is difficult and thus maintenance is limited
    b. The turbine must resist the force of tides and storms
    c. Ocean water conducts electricity easily.
    d. Power cables can become unburied, then damaged, by weather and by ships.

4. Any turbine fixed offshore must resist constant forces from tides and winds. Installing the foundation is usually the greatest cost of an offshore turbine.

5. Offshore wind turbines are difficult to maintain and repair. Therefore offshore wind turbines should be as maintenance free as possible.

6. Placing high voltage cables under water can be a dangerous decision. In order to eliminate this problem, best practices include:
   a. Place turbines away from recreational areas.
   b. Choose cable designs that are least likely to become damaged.
   c. Place the transformer on the shore rather than on a platform in the water.
   d. Use lower voltages to transmit electricity.

7. Placing cables under the ocean is a very important task. Note that laying the cables is one of the major costs for an offshore turbine (second only to installing the foundation for the turbine).

8. Underwater cables can become unburied. Therefore power cables must be buried to a sufficient depth, at least 3 meters below ground.

9. In general, the further out the wind turbine is from the coast the more power we can create. Therefore we should place wind turbines as far out as possible.

10. Placing an offshore turbine further into the ocean requires longer cables and an offshore transformer. The practical drawbacks are:
   a. More electrical hazards
   b. More locations for damaged equipment
   c. Difficulty in maintenance for the additional equipment

11. Placement of turbines must not interfere with shipping lanes and fishing areas. We must prevent wind turbines from being hit by ships. In addition, anchors and fishing nets can easily snap the cables. The best practices include:
   a. Avoid putting wind turbines in any shipping lane
   b. Put beacons on all turbines
   c. Bury cables to a sufficient depth

4.5
Additional Wind Power Terms

Introduction

When you examine reports on wind turbines you will often see certain technical terms. Most of these concepts I have described earlier, although I have not always used the technical terms. In addition, there are other technical terms that are mentioned quite often in manufacturer's data which I have not yet discussed. Therefore, in this chapter we will discuss additional terms and concepts which are important in understanding the technology of wind power.

List of topics for this chapter
1. Airfoil (blade) Design Terms
2. Aerodynamics, Speed, and Efficiencies

Airfoil (blade) Design Terms

1. airfoil: An airfoil is simply another term for the blade.

2. bottom of blade: (Fig. 4.4) The "bottom" of the blade is a side, not an edge. You can know the bottom of the blade by one of two ways: a) the bottom of the blade always faces into the wind, and b) the bottom of the blade is usually flat. It is the bottom side that is pushed.

3. top of blade: (Fig. 4.4) Like the bottom of the blade, the "top" of the blade is a side, not an edge. You can know the "top" of the blade by one of two ways: a) the top of the blade always faces away from the wind, and b) the top of the blade is usually rounded.

4. leading edge: (Fig. 4.6) The leading edge is the edge of the blade that faces into the wind. The wind splits around the leading edge, some wind going on the top of the blade, other wind passing under the bottom of the blade.

5. trailing edge: (Fig. 4.6) The trailing edge is the back edge of the blade. When the wind passes across the blade (either over the top of the blade or under the bottom) the wind travels from the leading edge of the blade to the trailing edge, then continues on its journey.

6. angle of attack: (Fig. 4.6) The angle of attack is the angle of tilt of the blade. It is the angle between the leading edge of the blade and the direction of the wind.

Figure 4.6 Blade Design Technical Terms

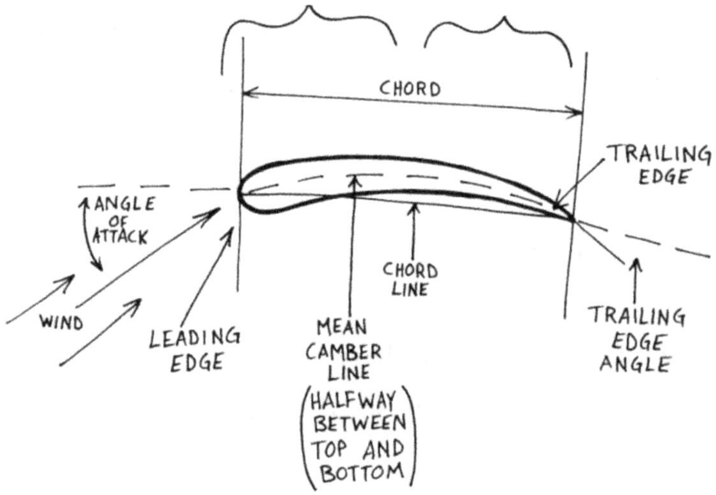

7. camber: (Fig. 4.6) As a verb, "camber" means "to curve." As a noun, "camber" means the amount of curvature. Modern wind turbine blades have a front edge that is curved. Therefore, blades are designed with a certain amount of camber.

8. camber line: (Fig. 4.6) The "camber line" is the geometrical line through the center of the blade. The camber line runs from the leading edge to the trailing edge.

9. chord line and chord: (Fig. 4.6) The "cord line" is the direct, shortest line from the leading edge to the trailing edge. The "chord" is the length of the chord line. (The chord is the shortest geometrical distance from the leading edge to the trailing edge).

10. <u>HAWT</u>: Horizontal Axis Wind Turbine

In a HAWT the axis of the rotor is parallel to the ground (and hence "horizontal.") These are the most common wind turbines.

11. <u>VAWT</u>: Vertical Axis Wind Turbine

In a VAWT the axis of the rotor is vertical. VAWTs look like egg-beaters.

Aerodynamics, Speed, and Efficiencies

12. <u>lift force</u>: (Fig. 4.4) The lift force is defined as the force perpendicular to the direction of the wind. It is a result of unequal forces on the bottom and top of the blade.

13. <u>drag force</u>: (Fig. 4.4) The drag force is parallel to the direction of the wind. The causes are numerous, and beyond the scope of this book.

14. <u>stall</u>: It is important to realize that stall is not related to the engine, but rather stall is related to the blade's angle of attack (and ultimately to aerodynamic lift).

When people associated with wind power or airplanes speak of a "stall" they are usually referring to a consequence of the blade's angle of attack. When a stall occurs, the angle of the blade relative to the wind is too large, therefore the lift force is reduced, and thus aerodynamic lift ceases to exist. When an airplane stalls, the plane is unable to fly. When a wind turbine stalls, the rotor slows down or stops altogether.

15. <u>tip speed ratio</u>: The "tip speed ratio" is defined as the speed of the tip of the blade as compared to the speed of the wind. Note that many turbine designers, even those who work with turbines for hydroelectric power, are concerned with tip speed ratio.

16. <u>cut-in speed</u>: The "cut-in speed" is the speed at which the turbine begins to produce power. In general, the choice for cut-in speed should be the lowest speed of the most common wind speeds in your location. Turbine manufacturers usually cite their cut-in speed at 10 mile/hour or more.

17. <u>cut-out speed</u>: The "cut-out speed" is the speed at which the turbine automatically shuts off. This is also known as the over-speed value. This is the wind speed value that is too much for the turbine to handle.

18. <u>rated wind speed</u>: The "rated wind speed" is the lowest wind speed at which your turbine should create maximum power. At the rated wind speed (and higher) the efficiency of the turbine is near 100%. The rated wind speed value is a claim made by the manufacturer.

19: <u>Reynolds Number, Re</u>: The "Reynolds Number" is a way of telling how well a fluid will move past an object. In regards to wind turbines, the "fluid" is the wind, and the "object in the way" is the turbine blade. A higher Reynolds number is an indication of a better turbine. Note that the Reynolds Number is a ratio of forces:

$$Re = \frac{\text{Force of the wind trying to get past blade}}{\text{Force of wind interacting with blade and thus not going forward}}$$

In practical terms this ratio is essentially: the force of the wind going forward versus the force of the wind *not* going forward. For a typical wind turbine, the Reynolds number will be between 500,000 and 10,000,000. Remember that Re is a ratio so there are no units. These numbers are often abbreviated with E6, which means "$\times 10^6$." Thus, "Re= 2.5 E6" actually means that the Reynolds value is 2.5 $\times 10^6$, or 2,500,000.

20. <u>Debris hitting blade</u>
    When debris hits the turbine blade the blade may become damaged. The pock marks and scratches may seem small to us, however these pock marks will reduce the turbine speed and therefore reduce the turbine efficiency. The material of the blade should therefore be chosen to resist any flying debris. This issue is most important where wind turbines are designed to provide a major source of electricity to a large region.

Conclusion

Many Americans hold passionate views about electrical power, yet few Americans understand all the details behind their passion. Electricity should not be mysterious. The science, the technology, and the data of electrical power can be understood by anyone.

Above all else, we must remember that there are no perfect solutions, there are only choices. Any option can be beneficial, yet each option has its own technical issues to work with. It is up to you and to your community to make those educated decisions. I hope that this book will help guide you in your choices.

M.F.

Obstruction Angles of Wind Turbines

Turbine Height:	100 ft.	200 ft.	300 ft.	400 ft.
Distance	Visual Angle	Visual Angle	Visual Angle	Visual Angle
1 ft	89.4°	89.7°	89.8°	89.8°
10 ft	84.3°	87.1°	88.1°	88.5°
100 ft	45.0°	63.4°	71.5°	75.9°
250 ft	21.8°	38.6°	50.2°	57.9°
500 ft	11.3°	21.8°	30.9°	38.6°
750 ft	8.1°	14.9°	21.8°	27.9°
1,000 ft	5.7°	11.3°	16.7°	21.8°
1/4 mile	4.3°	8.6°	12.8°	16.8°
1/2 mile	2.1°	4.3°	6.48°	8.6°
1.0 mile	1.1°	2.2°	3.2°	4.3°
2 miles	.54°	1.1°	1.6°	2.1°
3 miles	.36°	.72°	1.1°	1.4°
4 miles	.27°	.54°	.81°	1.1°
5 miles	.22°	.43°	.65°	.87°
6 miles	.18°	.36°	.54°	.72°

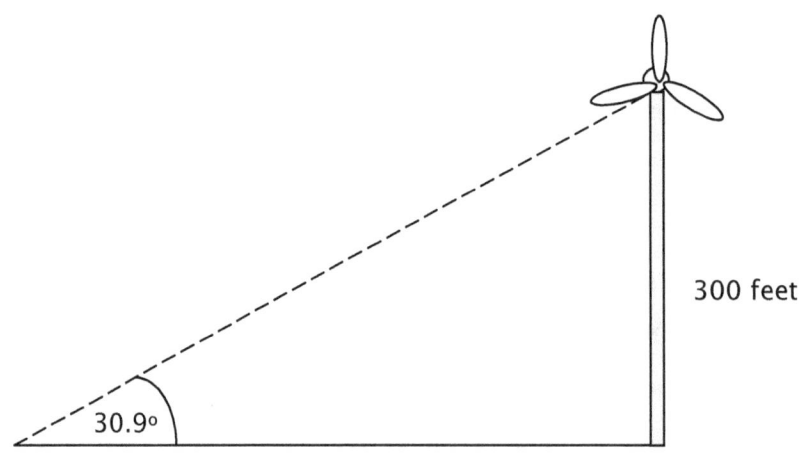

300 feet

30.9°

500 feet

Bibliography

Wind Power

1. <u>Energy for Man: From Windmills to Nuclear Power</u>, by Hans Thirring, 1958. Publisher: Indiana University Press.
2. <u>Energy Resources</u>, by Andrew Simon, 1975. Publisher: Pergamon Press, Inc
3. <u>Nontechnical Guide to Energy Resources</u>, by Ben Ebenhack, 1995. Publisher: PennWell Publishing Company
4. <u>Electric Power Generation: A Nontechnical Guide</u>, by Barnett and Bjornsgaard, 2000. Publisher: PennWell Publishing Company
5. <u>Power Surge: Guide to the Coming Energy Revolution</u>, by Flavin and Lenssen, 1994. Publisher: W.W. Norton & Company
6. <u>Energy: A Guidebook</u>, by Janet Ramage, 1997. Oxford University Press.
7. <u>Wind and Water Power</u>, by Clint Twist, 1993. Publisher: Gloucester Press
8. <u>Wind Energy Basics: A Guide to Small and Micro Wind Systems</u>, by Paul Gipe, 1999. Chelsea Green Publishing Company
9. <u>Wind Energy in America: A History</u>, by Robert Righter, 1996. University of Oklahoma Press
10. <u>Wind Power for the Homeowner</u>, by Donald Marier, 1981. Rodale Press
11. <u>American Wind Energy Association</u>, www.awea.org
12. <u>Wind Energy Explained: Theory, Design, and Application</u>, by Manwell, McGowan, and Rogers, 2002. Publisher: John Wiley & Sons, Ltd.
13. <u>Putnam's Power from the Wind</u>, Second Edition, by Gerald Koeppl, 1982. Publisher: Van Nostrand Reinhold Company
14. <u>Airfoil Diagrams</u>, by Bill Beaty http://amasci.com/wing/airgif2.html
15. "Catching the Wind" by Jim Motavalli, *E: Environmental Magazine*, Jan 2005.
16. <u>Danish Wind Power Association</u>, www.windpower.org/en/core.htm
17. <u>European Wind Energy Association</u>, www.ewea.org
18. <u>British Wind Energy Association</u>, www.bwea.com
19. <u>Irish Wind Power Association,</u> www.iwea.com/index1.html
20. <u>Airtricity Wind Turbines</u>, www.airtricity.com
21. <u>Bonus Wind Turbines</u>, www.bonus.dk/uk/index.html
22. <u>GE Wind Turbines</u>, www.gepower.com/businesses/ge_wind_energy

Government Sites – General

1. US Department of Energy (DOE) www.energy.gov
2. US Department of the Interior www.doi.gov
3. US Bureau of Reclamation www.usbr.gov
4. US Department of Agriculture (USDA) www.usda.gov
5. Environmental Protection Agency (EPA) www.epa.gov
6. Food and Drug Administration (FDA) www.cfsan.fda.gov
7. National Institute for Occupational Safety and Health (NIOSH) www.cdc.gov/niosh
8. Mine Safety and Health Administration (MSHA) www.msha.gov
9. Federal Energy Regulatory Commission (FERC) www.ferc.gov
10. Nuclear Regulatory Commission (NRC) www.nrc.gov
11. National Climatic Data Center (NCDC) www.ncdc.noaa.gov

Department of Energy (DOE) Related Sites

1. Department of Energy (DOE) www.energy.gov
2. Energy Information Administration (EIA) www.eia.doe.gov
3. [Office of] Efficiency and Renewable Energy (EERE) www.eere.energy.gov
4. Office of Fossil Energy (in Dept of Energy) www.fossil.energy.gov
5. Electric Transmission and Distribution Office www.electricity.doe.gov
6. Science (Office of Science) www.sc.doe.gov
7. Nuclear Regulatory Commission (NRC) www.nrc.gov
8. Civilian Radioactive Waste Management (OCRWM) www.ocrwm.doe.gov
9. Yucca Mountain Project www.ocrwm.doe.gov/ymp/about/index.shtml
10. International Nuclear Safety Program http://insp.pnl.gov
11. International Nuclear Safety Center, Argonne Laboratory www.insc.anl.gov
12. National Energy Technology Laboratory (NETL) www.netl.doe.gov
13. National Renewable Energy Laboratory (NREL) www.nrel.gov
14. Oak Ridge National Laboratory www.ornl.gov
15. Los Alamos National Laboratory (LANL) www.lanl.gov/worldview
16. Pacific Northwest National Laboratory (PNL) www.pnl.gov
17. Starlight, from PNNL/DOE http://starlight.pnl.gov

Energy & Environmental Research Center (EERC)

1. Energy & Environmental Research Center (at Univ. of North Dakota) www.undeerc.org
2. Coal Ash Resource Center www.undeerc.org/carrc/index.html
3. Center for Air Toxic Metals (CATM) www.undeerc.org/catm
4. Center for Biomass Utilization (CBU)
   www.undeerc.org/centersofexcellence/biomass/default.asp

www.ingramcontent.com/pod-product-compliance
Lightning Source LLC
Chambersburg PA
CBHW081357170526
45166CB00010B/3113